Face Detection and
Recognition on Mobile Devices

Face Detection and Recognition on Mobile Devices

Haowei Liu

AMSTERDAM • BOSTON • HEIDELBERG • LONDON
NEW YORK • OXFORD • PARIS • SAN DIEGO
SAN FRANCISCO • SINGAPORE • SYDNEY • TOKYO

Morgan Kaufmann is an imprint of Elsevier

Morgan Kaufmann is an imprint of Elsevier
225 Wyman Street, Waltham, MA 02451, USA

ISBN: 978-0-12-417045-2

British Library Cataloguing-in-Publication Data
A catalogue record for this book is available from the British Library

Library of Congress Cataloging-in-Publication Data
A catalog record for this book is available from the Library of Congress

For information on all MK publications
visit our website at www.mkp.com

Working together
to grow libraries in
developing countries

www.elsevier.com • www.bookaid.org

CONTENTS

Introduction to Computer Vision
on Mobile Devices

Computer vision, the field of how to enable computers to see the world, has been studied in the research community for a long time, and various technologies have been developed and are mature enough to be deployed in our daily lives. These technologies and applications include but are not limited to facial recognition for personal password login, moving object detection and tracking for video surveillance, and human activity recognition for entertainment purposes. These applications are typically made possible through one or multiple cameras mounted on stationary platforms. The backend software system—usually running on powerful machines such as personal computers or high-end servers—captures and analyzes the live video data and responds accordingly, depending on the target application.

In recent years, mobile computing devices such as tablets or smartphones have become more and more popular. Although these devices feature a low-power design and have less powerful computation capability as personal computers or mainframes, they still support most lightweight applications. Also, in addition to vision cameras, these computing devices usually come with sensors such as gyroscopes, accelerometers, pedometers, or GPS receivers, just to name a few. These sensors are typically not available on personal computers, and they open a new world on mobile devices for a wide variety of applications that are not seen on conventional platforms.

Researchers in academia and industry have started to look at developing computer vision algorithms on these devices to support and enable novel applications. On one hand, the applications could leverage the capabilities of additional sensors to assist the design of computer vision algorithms. On the other hand, the limited computing power and the interactive nature of mobile devices applications make it difficult to develop applications that are useful and natural. Targeting general engineers, these chapters will take a closer look at computer vision problems and algorithms and the possibilities of migrating them to mobile devices. It will cover the potential applications on these devices and encountered challenges. It will also introduce currently available platforms, software development kits, and hardware support. Specifically, the goals of this book include:

- To explain what computer vision is and why people would like to develop computer vision applications on mobile devices.
- To introduce commonly used state-of-the-art algorithms for different computer vision problems.
- To illustrate applications that might be made possible on mobile devices, and associated potential challenges.
- To describe available hardware, software, and platform support for developing mobile computer vision applications.

WHAT IS COMPUTER VISION?

Introduction to the Field of Computer Vision

Originated from the large field of artificial intelligence, computer vision is concerned with the study of enabling machines to perform visual tasks. The inputs to machines are images or video sequences, captured through a live camera feed. These tasks range from low-level de-nosing and filtering, to mid-level region growing or image/video segmentation (i.e., the operation of grouping image pixels into semantically uniform entities), and eventually to high-level detection or recognition (e.g., recognizing objects or faces in images). The old Chinese proverb, "A picture is worth a thousand words," best summarizes the challenges and opportunities of this field. While it is beneficial to have a vast amount of information readily available in such a small image, it is not an easy task to analyze the information, impose a structure on it, and reason about it in an efficient manner, although these tasks might be simple and intuitive for human beings.

The field of computer vision is closely related to many sibling fields such as signal processing, machine learning, or even cognitive science in psychology. For example, it leverages techniques in signal processing to remove the noise in the image data. It also uses machine-learning methodologies so that machines can "learn" to recognize objects, such as fruits. Understanding how human beings perceive the world can inspire new representations of image data and alternative solutions to some of the computer vision problems.

Figure 1.1 Example computer vision application—face detection/recognition. The left image shows an example where the camera automatically adjusts the focus based on detected faces. The right image shows an example of auto login with recognized faces.

What Can Computer Vision Do for Us?

After several decades of research and study, commercial products built on computer vision technology are already available in the market. For example, the built-in face detection chip inside digital cameras allows a camera to adjust the focus and exposure to the best quality automatically. By detecting the face regions and using the landmark information in the face (e.g., eye locations), a camera is also able to detect blinking eyes, red-eye effects, or smiling and signal the users accordingly. Some applications go one step further by recognizing the faces. The recognized faces are used as substitutes for usernames and passwords so the users can log onto their PCs without going through the irritating process of typing a lengthy password. Figure 1.1 shows example snapshots of these two scenarios.

Another area that heavily relies on computer vision technologies is the security or video surveillance domain. For instance, by mounting one or more cameras overseeing the parking lot or the entrance area to a building, the backend computer running computer vision algorithms could analyze the video, detecting and tracking any moving objects. The technology could provide functionalities such as zone intrusion detection or trip wire to the users (e.g., allow the security personnel to define a zone for any moving object detections). The resulting trajectories, along with object meta-information—such as moving speed, object classes (e.g., person or car), or color—can be stored in a database for future retrieval, which is useful to support queries such as, "Find me all the red cars that traveled greater than 30 mph yesterday morning" (Figure 1.2).

Figure 1.2 Example computer vision application—video surveillance. The left shows an example where the computer vision system tracks any moving objects and classifies them into different categories, such as cars or human beings. The computer vision system can also be used for indoor surveillance, e.g., detecting suspicious activities such as an unattended bag in train stations or office areas. More details could be found at the PETS project website: http://www.anc.ed.ac.uk/demos/tracker/pets2001.html.

Aside from the aforementioned applications, retail stores also adopt computer vision systems to prevent self-checkout loss or sweethearting (i.e., cashiers giving away merchandise to customers). By analyzing the hand movements and combing object recognition results and the barcode information, these systems are able to detect false-scans and report suspect actions and items to the storeowners. As an example, the Samsung Smart TV recognizes the hand gestures with a built-in camera to enable the users to control the TV with hand gestures. Yet another example would be the Kinect/X360, where by analyzing the depth images, the software could keep track of different parts of the human body and recognize actions so that the players can interact with the machine and enjoy the game in an immersive way. The availability of the Kinect sensor also triggers a massive amount of interest in creating systems that provide gesture-control features, realizing the vision depicted in the movie *Minority Report*, where Chief John Anderton (acted by Tom Cruise) interacts with the mainframe computer with hand gestures.

WHY MOBILE PLATFORM?
What Do We Mean by "Mobile"?
With the emerging of more advanced semiconductor technologies and wider deployment of internetworking devices, our daily lives are full of everyday objects and devices, capable of achieving ubiquitous or *pervasive computing*, a term describing the phenomenon of enabling and integrating information processing or computing capabilities into every

ordinary object surrounding human beings. Examples include Internet-enabled refrigerators, which make it possible for housewives to query and store recipes, or living room lamps that can be controlled remotely. These intelligent gadgets are made possible through embedded chips inside larger appliances that have moderate computing power.

In these chapters, we focus on platforms that are mobile—i.e., devices that are carried around and are used anywhere and anytime by the users. Specifically, we focus on smartphones and tablets, which provide a rich set of interaction features and a certain degree of computing power. Although laptops can be carried around as well, we exclude them from the discussion due to their similarity with desktops in terms of features provided. In the remainder of this section, we will provide the motivation as to why we decide to focus this book on emerging mobile platforms. In particular, we will show the sales numbers and market share growth to justify our motivations.

Mobile Devices Markets

Since the first intelligent phone went onto the market, the sales of smartphones have been rapidly increasing. Figure 1.3 shows the historical sales numbers of PCs, smartphones, and tablets.[1] We can see that the sales number of smartphones surpasses that of PCs, beginning in 2011, and it also maintains a steady growth. In addition, with the tablet computers gaining more popularity, tablet sales are forecasted to be much more than PCs and sell over 300 million units by 2016.

With these numbers, every major player in the technology sector would likely want to get into the highly lucrative and rapidly growing mobile markets. Currently, in terms of operating systems or mobile platforms, Google's Android and iOS from Apple take a majority share of the mobile market (Figure 1.4). Samsung and Nokia take the lead in terms of mobile devices shipped worldwide.

Applications or Apps on Mobile Devices

With the abundance of these nonconventional mobile devices, software engineers and application developers grab the opportunities to develop lightweight applications, or Apps, for these platforms. Taking advantages of additional sensors (e.g., gyroscopic, pedometer, accelerometer, GPS receiver, or touch sensor), these mobile applications provide the

---

[1]Source: Gartner, www.gartner.com.

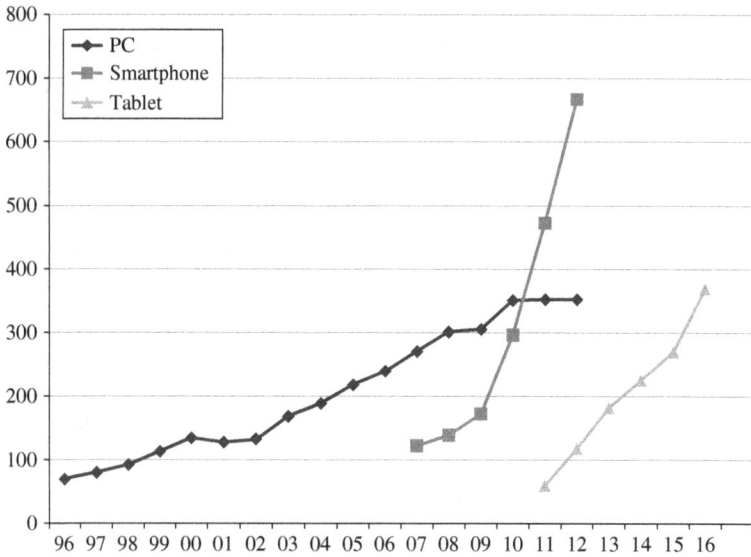

Figure 1.3 Yearly worldwide sales numbers for PCs, smartphones, and tablets (predicted numbers for tablets). Y-axis represents million units and X-axis represents years. The sales number of smartphones not only surpassed that of PCs but also quadrupled it in the past few years. The sales of tablets are also ramping up and are predicted to surpass those of PCs by 2016.

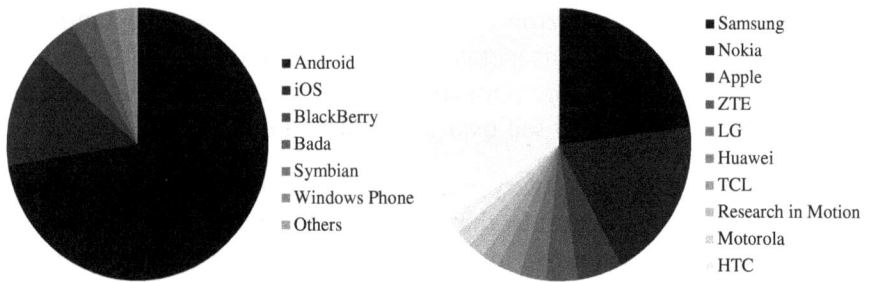

Figure 1.4 Global mobile market shares as of Q3 2012. The left figure shows the mobile market shares for each major player in terms of operating system. Google's Android takes about 70% of the market share, while iOS from Apple takes around 13%. The right figure shows the mobile market shares in terms of mobile devices shipped to the end users. Samsung and Nokia take about 40% of the market share. Gartner, www.gartner.com.

device users with new user experience and functionalities not available from software or applications on conventional platforms. For example, these novel applications include mobile games that allow the users to play games with touch gestures, such as Angry Birds or Fruit Ninja, personal health assistant applications that monitor the activity level of the device owner through a pedometer, or applications that plans trips with a GPS receiver.

The mobile applications are typically sold through an *app store*, a platform set up by operating system makers such as Google or Apple. Customers can download and use these applications by paying a small amount (<$10) through the platform. They can also rate the applications. A major difference of these mobile applications from conventional PC applications is that the users or nonprofessional engineers write most of them. This makes the software development cycle very short for these mobile applications but, at the same time, the rating system on the platform makes it easy for the application writers to receive feedback. The idea of writing software that many people can use and the low-cost nature of mobile applications drive the market growth and bring many more developers into the mobile world.

The popularity of mobile applications brings large revenues to the operating system makers as well, especially for Apple. As of late 2012, both Apple and Google have reached 25 billion downloads through the App Store and Google Play. The available applications for each are roughly around 700 k.[2] However, the average value of purchase from the App Store is much higher and brings Apple four times more revenues than what Google is making from its online store.[3]

COMBINING COMPUTER VISION WITH MOBILE COMPUTING

How does computer vision relate to these mobile devices, and what computer vision applications can be developed? What resources and challenges could the application developers leverage or encounter? In the next few sections, we will broadly touch on these issues.

Difference with Conventional Computer Vision Applications

While almost all mobile devices come with vision cameras, developing computer vision applications on these devices is different from so doing on PCs in the following aspects:

1. The application perspective is different. On PCs, the computer vision software takes a "third person" point of view. When analyzing the video feed, it first identifies the major subjects, who are

[2]http://news.cnet.com/8301-1035_3-57521252-94/can-apples-app-store-maintain-its-lead-over-google-play/.
[3]Source: AppAnnie.

either the interests of, or the ones that will interact with, the application. For instance, in video surveillance applications, the computer vision algorithms might identify moving targets (e.g., cars or people) while in the gesture recognition applications, the users would primarily be located. On the other hand, most mobile applications take the "first person" point of view. Through the camera, it sees what the users are seeing and hence, the focus of these applications shifts from the users to the scene or objects surrounding them.

2. The cameras are mounted in different ways. For mobile devices, the cameras usually move with the devices. This creates different motion artifacts for the applications. For example, hand shaking can blur the captured images, so camera motion would make video stabilization or frame correspondence necessary for high-level applications.

3. The availability of computing resources is different. Mobile devices usually have inferior computing power and less memory capacity compared with conventional platforms, which would put efficiency as an important design choice for application developers.

Challenges and Opportunities with "Going Mobile"

Given the aforementioned properties of these mobile devices, the challenges are different for the development of computer vision applications on these platforms. For one, because the devices are typically handheld by the users, it is crucial for the applications to reduce the artifacts caused by hand shaking during the imaging process. When analyzing videos, stabilizing the video frames or estimating the motion of the device to establish interframe object correspondences is also the key to building robust mobile computer vision applications. These problems are rare in conventional settings where cameras are mounted on stationary platforms.

Another critical issue is the computing resource. Mobile devices are designed for power efficiency, and therefore, these devices may not have as much computational resources as PCs. Hence, it is desirable when developing mobile vision applications to design them efficiently so that applications will not consume much battery life. However, for computer vision, it might not always be that straightforward to create computationally efficient applications. Some applications, such as gesture recognition, usually incur larger computation burden in order to achieve a reasonable robustness. One solution is to make a proper trade-off between robustness and power efficiency. Another possible solution is to off-load the computation to a remote server so the mobile devices are responsible for transmitting the data and interfacing the users.

Potential Impacts of Mobile for Computer Vision

The emergence of mobile devices makes them an important part of daily life. People use them to take pictures or videos of what they see, and then share and communicate with friends. Due to the change of perspective, the mobile applications are more customized to the users. Hence, these devices impact computer vision applications by shifting the focus to be more human-centric. For instance, the QR code reader applications recognize code from the captured images and allow the users to scan the code on any objects they see and show the web page or other relevant information to the users. The augment-reality applications fuse the surrounding of the users into a virtual world in a seamless way so they can interact with friends in an unprecedented manner. The gesture-control applications provide the users a new experience interacting with the devices, e.g., flipping a page with hand gestures when your hands are dirty.

SUMMARY

In this chapter, we started by explaining what computer vision is about, followed by example applications that computer vision might enable. We then moved to survey the current mobile market and provide numbers demonstrating the rapid growth of the mobile market. We also provided information regarding currently available mobile platforms. Finally, we provided motivations as to why developers would like to write computer vision software on mobile devices, and we discussed possible challenges and potential impacts. In the next chapter, we will touch on the technical details of developing mobile computer vision applications.

Face Technologies on Mobile Devices

Face-related problems have long been studied in the field of computer vision, and many technologies have also been developed in the industry. Given that the face is a unique exhibition of each individual, face-related technologies are mostly developed for surveillance and security applications and, recently, for entertainment purposes. For example, large-scale surveillance systems [1,2] are built based on detected facial features to allow law enforcement to query for videos of individuals with specific facial attributes such as being bald or wearing a beard. Face-recognition technology is widely used as well. For instance, the FastAccess software developed by SensibleVision allows mobile users to log on with their credentials without typing. The technology was also depicted in the movie *Mission Impossible: Ghost Protocol*, in which Agent Hanaway (played by Josh Holloway) uses his cell phone to detect the faces of the incoming people and identifies the target through a remote server. In addition, avatar animations controlled by human facial expression are becoming more popular, as are the built-in face detector or "smile detector" features in numerous cameras and mobile phones. Through face detection and recognition, it is possible that you could use an application on your mobile phone that would allow you to take photos of your friends, tag them, and upload the photos to Facebook. In this chapter, we will give a brief overview about the principles behind these technologies.

ALGORITHMS FOR FACE DETECTION
Overview of Face-Detection Algorithms
Creating reliable face detection has long been a problem, and solving it is a crucial step toward other facial analysis algorithms such as face recognition, tracking, or facial feature detection (e.g., eyes or nose). It is a challenging problem as different face poses, camera orientations, lighting conditions, and imaging quality could make the detection algorithms fail. With that said, much progress has been made in the last decade, and numerous approaches have been proposed and developed.

Yang et al. [3] extensively surveyed these algorithms and divided them into four categories:

1. The top-down, rule-based approaches detect faces using rules derived from the researchers' knowledge of faces. For example, the rules could be that the locations of the eyes have to be symmetric with respect to the center of the face or that the face region should be of uniform color. The drawback of a rule-based method is that it might not be that simple to derive precise rules from human knowledge. It is also difficult to extend the rules to cover all scenarios.
2. In contrast to the top-down approaches are the bottom-up, feature-based approaches. These approaches aim to find low-level features that signify the face region. They typically start with detecting features that could indicate face regions, such as regions with skin color, oriented edges that could indicate the lips, or textures that could indicate the cheeks. After the features are extracted, statistical models are built to describe their relationships and, later on, are followed by the verification of the existence of a face. Although these approaches enjoy a certain degree of success, they are sensitive to noise and occlusion. Shadows also have a big impact on the performance of a feature-based approach.
3. Template-matching approaches start by building a predefined template of the face. The template could be just an image of face silhouette or an edge map. It detects faces by correlating the template with the image, and regions that give the largest response are considered as face regions. However, a template-matching approach is inadequate, as it is not able to deal with variations in scale, poses, orientations, and shape effectively.
4. The appearances-based modeling approaches enjoy the most success. Different from the template-matching approaches, where the templates are handpicked by the researchers, an appearances-based approach attempts to "learn" a model from a large, labeled image dataset. The output of the learning process is a function the users can take to evaluate an image region. The function is called a *classifier* in the machine learning literature, and it will return whether or not the image region contains a face with a confidence value.

Many statistical models and machine-learning methods have been proposed for face detection, such as support vector machine, neural network, or Bayes classifier. However, the most widely used method is the one based on boosting, which we will touch on in the next section.

Viola–Jones Face-Detection Algorithms

The most popular and widely used face detection technique is the one developed by Viola and Jones [4]. It was developed as a generic object-detection framework and outperformed previously proposed approaches in terms of both performance and running time. The main challenge of face detection for these model-based or template-based approaches is the need to scan through every possible location and scale of an image in order to find the locations of the faces and reduce the false positives.

However, since an image usually contains only a few (0–5) faces, schemes like this end up spending lots of computing cycles on aggregating the information and evaluating candidates that are mostly non-face regions. The main contributions of the Viola–Jones face detector include a new image representation for fast feature computation, a boosting-base feature selection method, and a "cascade" classifier structure to quickly discard a large number of nonface regions. We will detail each of them next.

The Viola–Jones face detector starts with computing the features over a detection area on an intensity image. The features computed are rectangle features of different sizes. Three different kinds of rectangle features are used: two-rectangle features with two regions of the same size adjacent to each other horizontally and vertically. A three-rectangle feature consists of three neighboring rectangular regions, and a four-rectangle feature consists of two diagonal pairs of rectangular regions. For all cases, the feature value is computed as the difference between the sums of the pixel values within the black regions and the white regions. Figure 2.1 shows four examples of the rectangle features. You could think of these features as possible signatures for the face regions. The process of face detection is to look for the signatures, trim out nonface regions where the signatures do not exist, and identify the face regions with salient signatures.

In order to compute theses feature quickly, a new image representation, integral image, is introduced. Let the original image be I and let S be the integral image representation of I. The integral image S computes at each image location (x, y) a value $S(x, y)$, which is the sum of the pixel values above and to the left of (x, y), inclusively, on the original image. The computation of all the values in S can be done in a pass-through all the pixels on I following the recursion:

$$S(x, y) = S(x - 1, y) + S(x, y - 1) - S(x - 1, y - 1) + I(x, y) \quad (2.1)$$

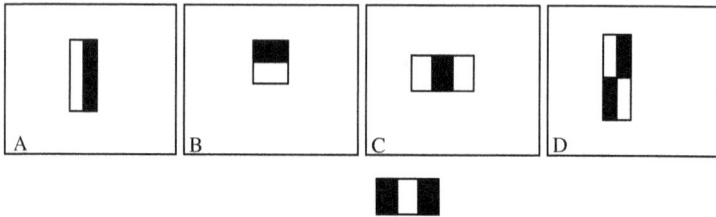

Figure 2.1 Rectangle features for face detection. The feature values are computed by subtracting the sum of the pixel values within the white regions from the black regions. A and B show two-rectangle features. C shows a three-rectangle feature and D a four-rectangle feature. Reproduced from [4].

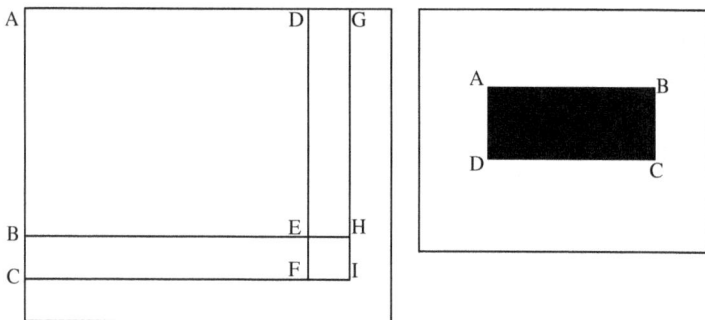

Figure 2.2 The left image shows the illustration of the recursion formula (2.1). The right image shows that to compute the sum of pixel values within rectangle ABCD on image I, we could look up the values on the integral image S to find out the cumulative sum at location A, B, C, and D. We can then compute the pixel sum on image I with S(C) − S(B) − S(D) + S(A), where S(·) indicates the value at a pixel location on image S. Reproduced from [4].

The recursion can be illustrated in the left image of Figure 2.2, which shows that the area of the rectangle ACIG can be computed by subtracting the area of rectangle ABED from the sum of those of rectangle ACFD and ABHG. Once the integral image representation S of the original image I is computed, the sum of original pixel values within any rectangle can be computed by table lookup. For example, as shown in the right image of Figure 2.2, the sum of the rectangle ABCD in I can be computed by $S(C) - S(B) - S(D) + S(A)$, where $S(\cdot)$ indicates the value of a particular location on image S. Note that, with the integral image, we could compute the sum of any rectangle region on image I using a constant number of operations, e.g., four lookup operations and three numerical operations. This allows us to compute the rectangle features in an efficient manner.

With the computed features, one can use this favorite classifier and train a face model for detection using these rectangle features.

However, with just a 24-by-24 detection window, there are about 160,000 possible rectangle features. It will be impractical and computationally infeasible to compute all of the features every time when we would like to know if a certain detection region contains a face. The question, then, is how to choose a small subset of features and build a classifier in an efficient manner.

Feature selection is a problem that has long been studied in the field of machine learning, and many approaches have been proposed. The Viola−Jones face detector uses a variant of boosting [5], as it could allow the classifier to efficiently prune out most of the irrelevant features during the training stage. Conventionally, boosting is a classification scheme used to boost the classification result of simpler learning algorithms. In the literature, the simpler learning algorithm is termed as a *weak learner*, meaning that its classification performance could be just slightly better than random guessing. For instance, for a two-class classification problem, the classification performance might be slightly over 50%. The boosting scheme adopts a round-based procedure and iteratively evaluates the weak learner on the training set. In each round, a set of weights is kept for the training set and the weight for the wrongly classified example is increased so that the weak learner could focus on learning the hard examples. At the end of each round, the weak learner is assigned a weight based on how well it does on the weighted training set. The better the learner performs, the larger the weight it is assigned. The output of the boosting procedure is then a weighted combination of these learners.

The boosting procedure can be seen as a process that produces a weighted sum of a large set of classification functions where strong-performing functions receive larger weights. It can also be used to incorporate the feature selection procedure if the weak learner is a function of a single feature. The Viola−Jones detector proposes the following function as the weak learner:

$$h(x, f, p, \theta) = 1 \text{ if } pf(x) < p\theta, \quad 0 \text{ otherwise} \qquad (2.2)$$

Here, h is the classification function that takes as parameters x, the image region within the detection window, f, the rectangle feature being considered, p, the polarity that determines the direction of the sign and, θ, the threshold for the feature and produces a decision regarding whether the detection window contains a face. Since the

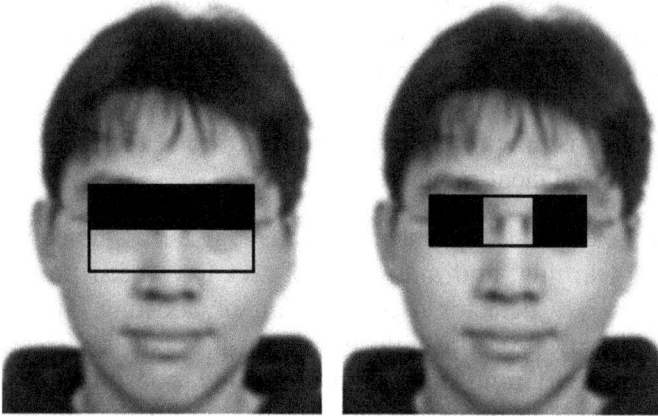

Figure 2.3 Reproduced from figure 5 in Ref. [4]. The rectangles shown on the frontal face indicate the first two features selected by the boosting procedure. The significance of these features can be interpreted based on the observation that the eye regions, which are usually darker than the cheek and the nose bridge in a typically frontal face.

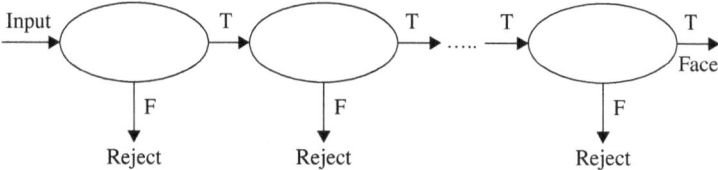

Figure 2.4 Depiction of the cascade classifier. The cascade classifier can be viewed as a decision tree. Given an input region, a series of classifiers are called upon to evaluate the input. A region has to go through every stage before it can be declared as a face region.

classification function h is a function of only one single feature, the boosting procedure amounts to selecting features that are more relevant. A key reason the Viola–Jones detector adopts boosting as a feature selection process is its efficiency as compared to other feature selection methods, which typically require an additional loop on the number of examples.

Figure 2.3 shows the first two features selected by the boosting procedure. These two features imply that a signature for a frontal face is a darker region (the eyes) or a sandwiched dark–light–dark pattern in the middle.

To further reduce the false positive rate and increase the detection accuracy, the Viola–Jones face-detection algorithm constructs a cascade of boosted classifiers as shown in Figure 2.4. The reasoning behind the cascaded classifier is that most of the false positives could be pruned out and rejected at the early stages. The classifiers at the earlier stages are tuned so that they could reject most of the false faces while at the same time maintaining a decent detection rate. Subsequent classifiers are trained using the positive examples that pass every previous stage and as a result, classifiers at later stages actually deal with harder examples. Interested readers could refer to Ref. [4] for more details.

At the time of publication, the method was running 15 frames per second on a Pentium III processor. However, even with the computing power of the modern day cell phones, it is still difficult for the original algorithm to run in real time on mobile platforms. We will introduce some modified face detection methods after we introduce face recognition algorithms.

ALGORITHMS FOR FACE RECOGNITION

Overview of Face-Recognition Algorithms

Given an input image and a detected face region, face recognition refers to the problem of identifying the detected face as one of those in a face database. The problem has been studied for over 30 years, and research has resulted in many successful commercial applications. However, two main challenges still remain: face pose variation and illumination changes.

Numerous approaches have been proposed due to the potential extensive applications. For example, in the security application domain, face recognition technology can allow law enforcement officers to search for a particular person in large video databases. In this section, we focus on image-based face recognition technologies and give an overview of different face recognition approaches.

The approaches to face recognition can be categorized into three classes [6]:

1. *Holistic approaches*: In this category, every pixel of the face region is fed into the recognition system. Two widely used and most successful face recognition algorithms, Eigenfaces [7] and Fisherface [8], belong to this category.
2. *Feature-based approaches*: In this category, the features or the landmarks on the face such as eyes, nose, and mouth are detected first.

Their locations and neighborhood are fed into a classifier or matched against a pretrained dataset for recognition.
3. *Hybrid approaches*: Approaches that combine the ideas from the two previous categories and potentially offer the best of both.

Over the past few decades, much progress has been made toward the detection of faces in still images and the extraction of the face features. At the same time, significant advances have also been made in the area of face recognition. Researchers have shown that holistic approaches are effective on large datasets. Feature-based approaches also enjoy high recognition rates, and compared with their holistic counterparts, they are shown to be more resistant to pose and lighting changes. However, their performance highly depends on the reliability of underlying feature detection and localization methods. For instance, the eye detectors might not work if they are based on the geometry model of open eyes. In the next few sections, we will introduce some representative methods from each genre, especially Eigenfaces [7] and Fisherfaces [8] due to their success and popularity.

Holistic Face-Recognition Algorithms—Eigenfaces and Fisherfaces

Given a query image, the main idea behind Eigenfaces is trying to find its nearest neighbor from the face dataset and then assigning the identity of the nearest neighbor to it. The question, then, is how to represent the face region. Assuming we could crop and scale the face region to an N-by-N square, we could denote the entire face region as x, where x is a feature vector of dimension N^2. Given a face dataset consisting of M faces, denoted as y_1, y_2, ... y_n, the Eigenfaces approach tries to find y_k, such that y_k minimizes the distance function $||\mathbf{x} - \mathbf{y_i}||$.

Note that x sits in a very high dimensional space. For example, if the face region were a 256-by-256 image patch, x would be a vector of 65,536 dimensions. Although the N-by-N dimensional space represents all possible images of size N-by-N, the actual distribution of face images might actually live in a subspace that has a much smaller dimension due to the similarity of different human faces. Figure 2.5 illustrates that the image dataset might sit in a lower-dimensional space, indicated by the solid ellipsoid.

To find such a lower-dimensional space, the Eigenfaces approach performs principal component analysis (PCA) on the image dataset.

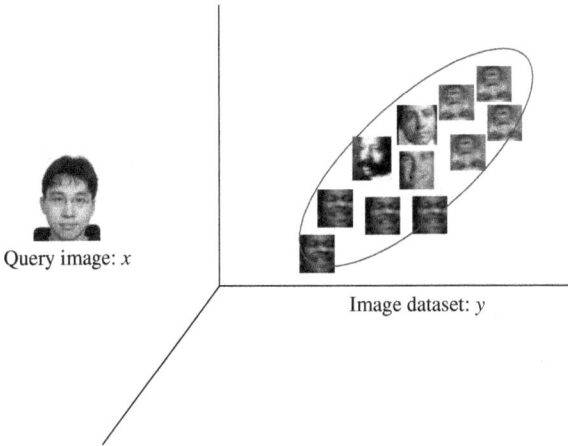

Query image: x

Image dataset: y

Figure 2.5 Illustration of Eigenfaces. Given a query image x, *the goal is to find the nearest neighbor in the face image dataset, sitting in a lower-dimensional space, indicated by the red ellipsoid.*

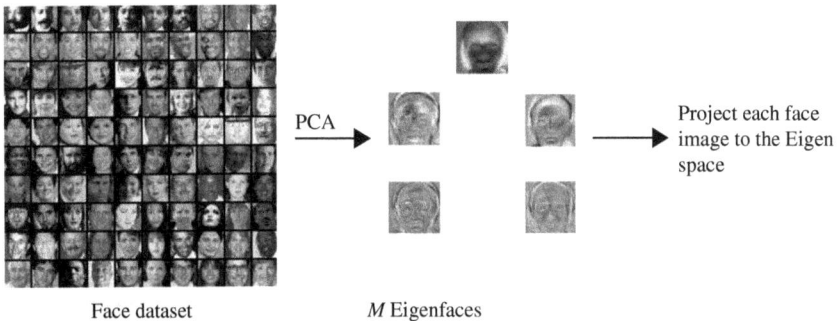

PCA

Project each face image to the Eigen space

Face dataset

M Eigenfaces

*Figure 2.6 Illustration of Eigenfaces process. Given a set of training face images, PCA is performed to find the top-*M *Eigenfaces. Each image in the dataset is then projected to the space formed by the Eigenfaces, i.e., each image is represented as a weighted sum of* M *Eigenfaces. These weights are the new representation of the original image in the new space. Given a query image, computing its nearest neighbor is done in the new space.*

PCA is a statistical method used to find the principal components or eigenvectors of sample data. These principal components characterize how the sample data are distributed. For instance, the principal components would be the variance along the x- and y-axis for data that are Gaussian distributed in a 2D space. In face recognition, these eigenvectors are called Eigenfaces and can be thought of as canonical faces that characterize the face image distribution. Figure 2.6 shows the process of using Eigenfaces for face recognition.

The Eigenfaces approach has been shown robust to noise such as partial occlusions, blurring, or background changes, and has yielded good performance on standard datasets. However, the drawback of Eigenfaces is its lack of discriminant power. The reason for this is that Eigenfaces do not take class information (i.e., face identities) into account. Instead, all the training images are fed into the PCA module at once with no class-specific treatment. Fisherfaces [8] was proposed to remedy such a drawback.

The motivation behind Fisherfaces is that since the training set is labeled with face identities, it will make more sense to use a class-specific linear method for both dimension reduction and recognition. Instead of PCA, Fisherfaces uses linear discriminant analysis (LDA) for dimension reduction. Given a set of face images with C classes, the Fisherfaces approach computes the between-class scatter matrix:

$$S_{\mathrm{B}} = \sum_{i=1}^{C} N_i(\mu_i - \mu)(\mu_i - \mu)^{\mathrm{T}} \tag{2.3}$$

and the within-class scatter matrix:

$$S_{\mathrm{W}} = \sum_{i=1}^{C} \sum_{x_k \in X_i} (\mu_i - x_k)(\mu_i - x_k)^{\mathrm{T}} \tag{2.4}$$

where N_i denotes the number of images within class X_i, μ the mean feature vector of all the images, and μ_i the mean feature vector of the images of class X_i. The Fisherfaces approach computes the eigenvectors W such that

$$W = \mathrm{argmax} \frac{|W^{\mathrm{T}} S_{\mathrm{B}} W|}{|W^{\mathrm{T}} S_{\mathrm{W}} W|} \tag{2.5}$$

The difference from PCA is that LDA maximizes the ratio of between-class scatter and the within-class scatter. The idea behind this is to project the feature vectors to a new space so that feature vectors from the same class are closer while those from different classes are further away. This can be seen in Figure 2.7. The solid circles and triangles represent two different classes in the 2D space. The dotted line is the subspace formed after PCA is performed, while the dashed line is that after LDA. Note that on the dotted line, the projected locations of the triangles and circles are mixed together, making it difficult to differentiate the two classes. On the other hand, on the dashed line, the

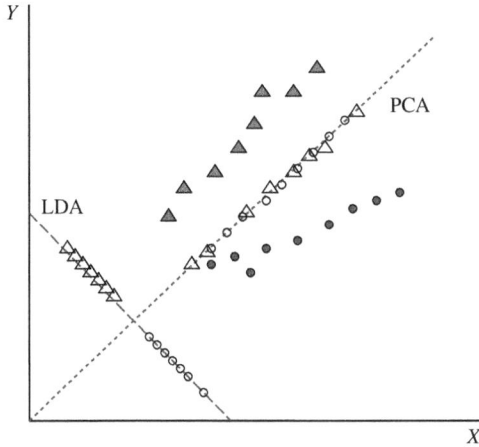

Figure 2.7 Comparison between PCA and LDA. The circles and triangles represent two different classes in a 2D space. The dotted line denotes the subspace formed after PCA and the dashed line denotes that after LDA. We can see that LDA produces a projection that is more discriminant.

projected circles are well separated from the triangles. This shows that LDA is more discriminant than PCA.

Other features have also been proposed, such as edges or intensities. Interested readers could refer to Ref. [6] for more details.

Feature-Based Face-Recognition Algorithms

Earlier face-recognition methods are mostly feature-based. For instance, they recognize faces using information such as the width of the head, distances between the eyes and from the eyes to the mouth, or the distances and angles between eye corners, mouth extrema, nostrils, chin, and so on. The most successful feature-based face-recognition algorithm is the elastic bunch graph matching system (EBGM) [9]. The approach represents the face regions with Gabor wavelets. That is, each image is convolved with the Gabor filter to generate wavelet coefficients:

$$g(x,y) = \exp\left(-\left[\frac{x^2}{2\sigma_x^2} + \frac{y^2}{2\sigma_y^2}\right] + 2\pi i[u_0 x + v_0 y]\right) \qquad (2.6)$$

where σ_x and σ_y denote the width of the spatial Gaussian and u_0 and v_0 denote the frequency of the complex sinusoid.

For each face in the training set, a set of N fiducial points, e.g., the pupils, corners of the mouth, and nose tip are labeled. A graph G is

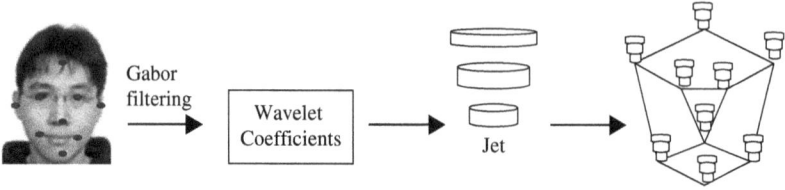

Figure 2.8 Illustration of EBGM. The image with labeled fiducial points is first applied Gabor filtering to generate wavelet coefficients. A jet (i.e., a set of wavelet coefficients at different scales and orientations) is extracted for each fiducial point. Finally, a graph representation is constructed for the face.

then constructed to represent the face, where each vertex of G corresponds to each fiducial. A feature representation, *jet*, is defined to describe the local image patch surrounding a particular image pixel. It is essentially a set of wavelet coefficients at different image orientations and scales. For each graph constructed, the jet for each fiducial point is also extracted. Given an input face image, the EBGM face recognition system extracts the fiducial points by matching it against an off-line built face model and represents it with a graph. Face recognition is done by finding the nearest neighbor of its graph representation. Figure 2.8 illustrates the process of graph building in the EBGM face recognition system. The elastic bunch graph matching approach has also been successfully applied to other domains, such as face detection, pose estimation, gender classification, and general object recognition.

Hybrid Face Recognition Algorithms

Finally, the hybrid approach combines the features from the previously mentioned genre. For instance, in Ref. [10], instead of computing the Eigenfaces for the entire dataset, the authors compute the Eigenfaces for each view point and show that it performs better. Taking the idea even further, they introduce the concept of Eigenfeatures, such as Eigeneyes and Eigenmouths and show that with a small number of eigenvectors, the Eigenfeatures approach outperforms Eigenfaces.

FACE TECHNOLOGIES AND APPLICATION ON MOBILE PLATFORMS

We have reviewed widely adopted face-related technologies in the previous sections. Although these algorithms enjoy much success and are popular on powerful computing platforms, it is difficult for them to run in real time on mobile platforms given the limited computing resources. In this section, we will look at approaches that adapt these algorithms to mobile platforms and the applications.

Face Detection on Mobile Platforms

A face detector based on the cascaded classifiers runs efficiently on a PC but would need further optimization to run on mobile platforms. Although there are dedicated hardware implementations—e.g., smart digital cameras for face detection—real-time software implementation of face detection is still a challenge. Several strategies have been proposed to address this issue, categorized as follows:

1. *Limited application scenario*: Users are mostly interested in mobile application related to frontal face only. This simplifies the face-detection problem since it implies that developers could ignore face detectors for profile faces and only the one for frontal face is required.
2. *Spatial subsampling*: The most intuitive and the first attempt to reduce the required computations is reducing the resolution of the input image. However, arbitrarily scaling down the input image may cause the performance of face detectors to degrade.
3. *Detector parameter tuning*: Another strategy is to fine-tune the detector parameters. For example, in Ref. [11], the authors scale up the detection window and the step size in order to reduce the number of scanned locations in the image.
4. *Hardware support*: Software written on mobile devices may be able to use the hardware resources to improve the run-time performance. For example, using SIMD (single instruction and multiple data) instructions, GPU, or fixed-point operations instead of floating-pointing operation can all improve the frame rate [11].
5. *Tracking:* An effective way to reduce computation is through tracking, which uses the information from the previous frames to help identify the face locations in the current frame. The assumption is that in typical applications, the face will not move much from frame to frame. In Ref. [11], the authors identify a keyframe every 30 frames and the full face detection pipeline is only run on the keyframe. For the rest of the frames, face detection is only run on the surrounding region of the face location from the previous frame, and thus avoids the need to search for faces at nearly every location in the current frame.
6. *Additional sensor:* One drawback of the Viola—Jones face detector is that the feature is not orientation invariant. To detect faces of different orientations, the training set has to contain example images

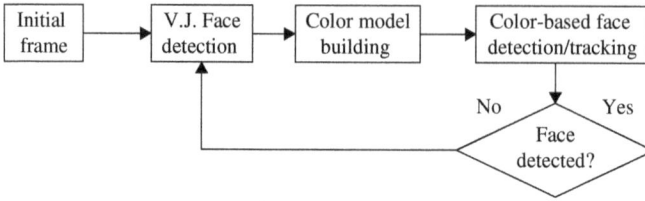

Figure 2.9 Procedure of the hybrid face tracker. V.J stands for Viola-Jones.

or a classifier has to be trained for each orientation. In Ref. [12], the authors use the gyroscope on the mobile phone to identify the orientation of the handheld device and rotate the image accordingly.

7. *Using auxiliary information:* Another effective way is using auxiliary information to find only a few possible face candidates and using the face detector to verify those candidates. For instance, in Ref. [13], the authors propose a hybrid face detection and tracking procedure that goes even beyond the keyframe approach. The full cascaded classifier is only run on the calibration frame to find the location of the faces. A skin color model is then built using the detected face regions. Later on, the detected face is tracked using the skin color model and face detection is applied on the region that surrounds the skin mask for verification purposes. If the color-based tracker fails, a full search using the face detector is performed to reinitialize the face location. The resulting frame rate is about 20 frames per second on VGA images (image with size 480 by 640 pixels). Figure 2.9 shows the pipeline of the hybrid face tracker. A similar approach is presented in Ref. [14] as well.

Face-Detection Applications on Mobile Platforms

Face detection is usually the first step towards many face-related technologies, such as face recognition or verification. However, face detection itself can have very useful applications. The most successful applications of face detection would probably be photo taking. When you take a photo of your friends, the face detection algorithm built into your digital camera detects where the faces are and adjusts the focus accordingly.

Head pose estimation is another application that heavily relies on face detection. Estimating the head pose is useful in the settings of automated guided cars, where an in-car device runs the head pose estimation algorithm to detect the drowsiness of the driver. In Ref. [15],

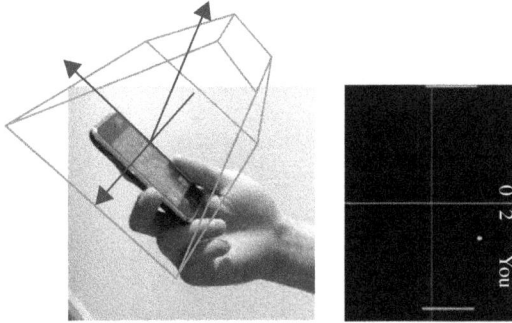

Figure 2.10 Reproduced from the figures in Ref. [16]. The left image shows the concept of MIXIS based on face tracking. The right image shows a possible Pong game application enabled by face tracking.

the authors show a real-time head pose estimation system that can identify five poses on a mobile platform.

Another good usage of face detection/tracking is the support of mobile device interaction. In Ref. [16], the authors propose an inter-action method with the mobile device based on tracked face position. With mixed interaction spaces (MIXIS), their system allows the mobile device to track the location of the device in relation to a tracked feature in the real world. The main idea of MIXIS is to use the space and information around the mobile device as input instead of limiting interaction with mobile devices to the device itself (Figure 2.10). To this end, they track the location of the user's face to estimate the location of the device relative to the tracked face, which forms a 4D input vector, i.e., x, y, z, and tilt. The difference with other face-tracking applications is that in their system, it is not the face that moves but the mobile device, which minimizes exhaus-tion related to frequent head movements. With the face location tracked, the system can support applications such as a Pong game or map panning and zooming. Figure 2.10 shows the concept of MIXIS, where the user is able to interact with the mobile device by moving it in different directions. This is achieved through the face-tracking module on the device. The image on the right-hand side shows a Pong game application.

Face Recognition and Verification on Mobile Platforms

Compared to face detection, the algorithms for face recognition typi-cally run much faster. Therefore, with a fast face-detection implemen-tation, face recognition can typically run on mobile platforms without

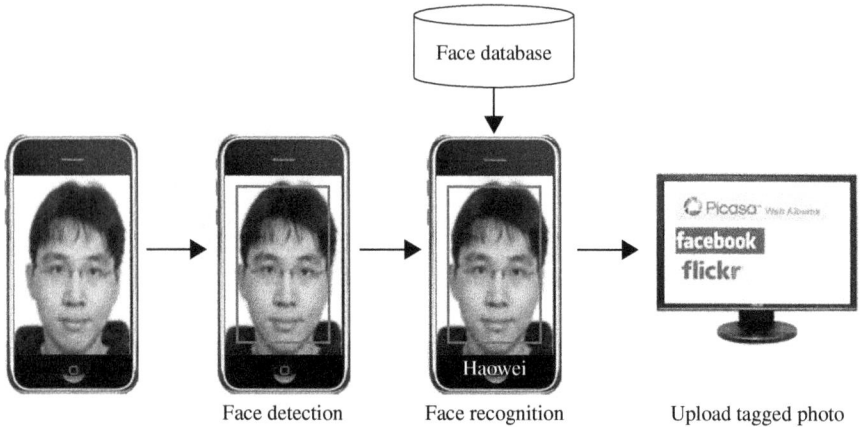

| Face detection | Face recognition | Upload tagged photo |

Figure 2.11 Reproduced from of the figures in Ref. [17]. Given an input photo taken by the user, the system first identifies the face region. It then uses an eye detector to detect the landmarks on the face. The landmarks are used to rotate the image so it is aligned with the orientations of those in the face database. Finally, the system extracts features and classifies the face into one of those in the database.

many modifications. With that said, face recognition enables many novel applications on mobile platforms. In Ref. [17], the authors present a system for face annotation. The user takes a photo using a mobile phone and the system detects the faces in the photo, recognizes and identifies the faces, and finally posts the photo with annotated faces to the user's social network websites. Figure 2.11 shows a high-level illustration of this system.

The system was implemented on the Android platform on the Nvidia Tegra SoC board. To start with, it detects the face region using the Android API. It then builds a custom Ada-boost eye detector to identify the landmarks on the face. Given the landmark locations, the system scales and registers the input photo so the image is better aligned with those in the database. Later, the system extracts Gabor wavelet features [18] on the face region and dimensionally reduces using PCA and LDA. Finally, the system uses a K-nearest neighbor classifier for face recognition.

Although the Gabor wavelet features are reported to be more invariant to illumination and pose changes, they are more computationally expensive. On an ARM CPU, the authors report a 5.1 s per-frame performance. Using an embedded GPU, the authors are able to reduce the computational cost to one second. The entire system achieves over 90% on a traditional face dataset.

140	141	128
136	139	144
137	140	143

1	1	0
0		1
0	1	1

Figure 2.12 LBP of the centering pixel. To compute the LBP for the centering pixel (the one with pixel value 139), we go through its eight neighbors. We mark one on the right if the pixel value of the neighboring pixel is larger than 139 and 0 otherwise. The binary numbers on the right indicate the LBP of the centering pixel.

Another major application is to use face authentication to unlock the mobile phone [19]. That is, instead of asking the users to type or gesture, the mobile phone can unlock itself by recognizing the face of its user. Similar ideas have been implemented, such as the face unlock feature on Android, or other mobile platforms [19,20]. The pipeline for face authentication is similar to that in Figure 2.11, except that the number of stored faces is much smaller, as frequent users of a mobile phone are typically limited to one or two. The feature extraction part could differ to accommodate for different applications. Next, we briefly discuss some examples and considerations for feature design.

In Ref. [19], the authors propose to use local binary pattern (LBP) [21] for face recognition on the mobile platform for its computational efficiency and discriminant power. An LBP is a representation of a pixel on an image. For a pixel, its LBP can be formed by comparing the pixel to each of its eight neighbors. When its pixel value is greater than the neighbor's value, mark "1," otherwise, mark "0." This gives an 8-digit binary number for a pixel and is called the LBP for the pixel. Figure 2.12 shows an example of LBP for the center pixel.

To use LBP as the feature for facial authentication, the authors first divide the detected face region into several grids. The LBP is computed for each pixel in each grid. The binary number can be turned into a decimal value. For instance, the binary number 00001000 is eight in decimal base. With these values, the authors compute a histogram for each cell and use the concatenated histogram to represent the detected face. The concatenated histogram is compared to that of the stored face to determine if there is a match. Figure 2.13 shows the process of using LBP for face authentication. The authors later augment the authentication system by combining face and speech recognition. Interested readers could refer to Ref. [22].

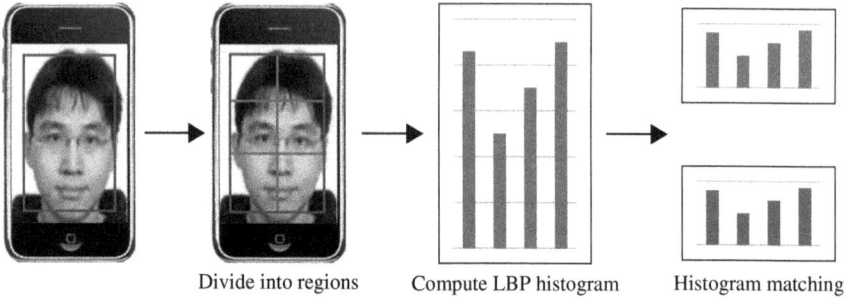

Divide into regions   Compute LBP histogram   Histogram matching

Figure 2.13 Pipeline for using LBP for face authentication. Given an input image with a detected face region, the process first divides the face region into regular grids. An LBP histogram is constructed for each grid. The final concatenated histogram is formed as the feature for the face. It is matched with the histogram of the stored face to authenticate the input face.

Another face verification system is presented in Ref. [20]. The authors use template matching in the frequency domain to determine if the query image is authentic. During the training stage, the face regions are detected and extracted from the training images. These detected regions are scaled to ensure all faces from the training set have the same dimension. Finally, a template is constructed from the training set. In addition to a global face template, to improve the accuracy, the authors extract subregions around facial features, such as eyes and nose, and build several part templates. Given a query image, the system matches it with the global template while at the same time extracts the regions around the facial features to produce several fragments, each of which is matched with the part templates to produce a local matching score. The global and local matching results are combined to determine if the query image is authentic. In all cases, fast Fourier transform is applied to all the images and the correlation is done in the frequency domain to improve computational efficiency. Figure 2.14 illustrates the process of the system.

Facial Feature Tracking on Mobile Platforms
The Active Appearance Model
So far, we have been discussing the use of facial features, such as eyes and nose tip, for facial recognition or verification. However, being able to track the facial features reliably over time is itself useful. For instance, by recognizing and tracking the facial features, you could imagine a videoconferencing application where an avatar speaks on your behalf and its facial geometry morphs as you speak or move your head. Figure 2.15 shows a videoconferencing system where the software

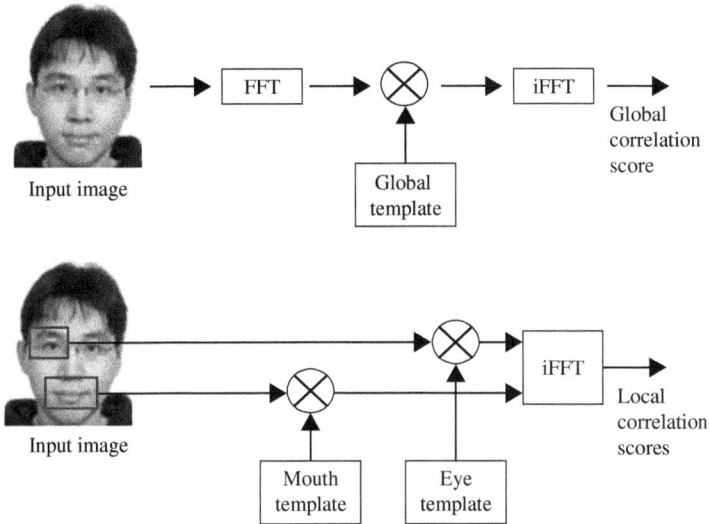

Figure 2.14 Illustration of the pipeline in Ref. [20]. Each input image is not only globally matched with a prebuilt template (top) but also partially matched with several part templates (bottom). The two correlation scores are combined to determine if the input image is authentic.

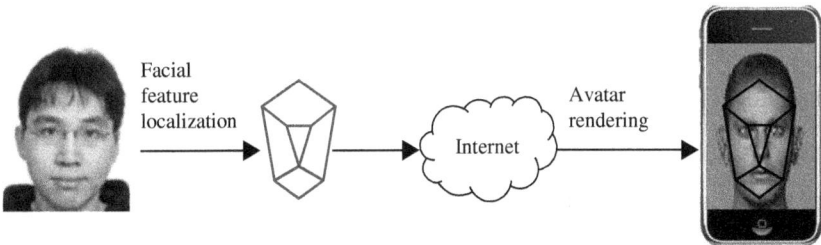

Figure 2.15 Illustration of the mobile videoconferencing application. In a mobile environment, the network bandwidth is limited and hence it is not possible to use conventional videoconferencing software that sends all the image data over the network. With the facial feature tracker, the application on one end could just send the landmark locations over, and upon receiving the information, the application on the other end could control the face geometry of the avatar accordingly.

on one end tracks the facial features and sends their location information over the Internet. The motivation behind this is that in a mobile environment, the network bandwidth is limited, and hence it is not possible to use conventional videoconferencing software that sends all the image data over the network. With a facial feature tracker, the application on one end could just send the landmark locations over and the application on the other end could render an avatar and control the avatar with the received facial information.

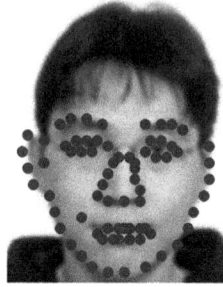

Figure 2.16 An example of hand-labeled image with landmark points. Red dots indicate the locations of landmarks of interest. Reproduced from figure 1 from Ref. [23].

As another example, imagine applications that use the facial feature information to infer where the users are looking and then render the user interface accordingly. The availability of the detailed feature information could also enable finer-grained interface control than using the head pose, e.g., allowing disabled users to control the interface by eye blinking. It can also be applied to authentication, as we have discussed previously.

In this subsection, we introduce the state-of-the-art method for facial feature tracking, and how it can be optimized on mobile platforms in the next subsection. We conclude the section with its applications, particularly in the area of authentication.

The conventional approach toward this end is the active appearance model (AAM) [23]. The authors propose this as a means to interpret an image. Their approach interprets an unseen image by matching it against a synthesized or rendered image from an offline model. The approach falls in the category of what is called analysis-by-synthesis in the literature. For example, in order to analyze face images, the authors generate a statistical model for face shape variation. The model is generated offline with a set of face images, and each image contains n hand-labeled landmark points. Figure 2.16 shows an example of a hand-labeled face image. The red dots represent the landmark points.

For a labeled face image with n landmark points, we could represent its shape with a $2n$ dimensional feature vector $X = (x_1, y_1, x_2, y_2, \ldots x_n, y_n)$, where x_i, y_i is the image coordinate of landmark point i. Given a set of N aligned labeled images, the AAM approach computes the mean vector $\overline{X}$ as the mean shape and models the shape of each face X as $\overline{X} + \delta$, where δ is the shape variation.

AAM assumes δ is Gaussian distributed and applies PCA to find the canonical variations. Finally, the shape of each face can be written as $\overline{X} + Pb$, where P is the matrix whose columns are the top-k eigenvectors from PCA and b is the k dimensional shape parameter.

The above procedure only models the shape variations of the face. To model the appearance, the AAM approach first warps each example image so that it aligns with the mean shape. For each example face image, let g be the vector representing the intensities of the pixels covered by the mean shape and $\overline{g}$ be the mean appearance vector of all the images. Similarly, the appearance of the face could also be modeled as $g = \overline{g} + \delta_g$, where δ_g represents the appearance variations and, in turn, can be written as $\delta_g = P_g b_g$, where P_g is the matrix whose columns are the top-k eigenvectors from PCA done on the appearance feature vectors and b_g is the k dimensional appearance parameter. Figure 2.17 shows examples of synthesized images using the AAM model.

Given a statistical face model and an unseen image, how can we then synthesis a face image that matches the unseen image well in order to analyze it? Using a face detector, the AAM approach translates and scales the face region to approximately align it with the face model. Let $c = (b, b_g)$ be the model parameters from which a model image I_m is generated from and I_i be the input image. The matching problem can be formulated as finding the optimal model c, such that the $|\delta I|^2 = |I_{m-I_i}|^2$ difference between the model image and the input image is minimized, which is a nonlinear high-dimensional optimization problem. To find the best model parameters c, the authors propose to learn the mapping between δI and δc. The assumption is that the spatial

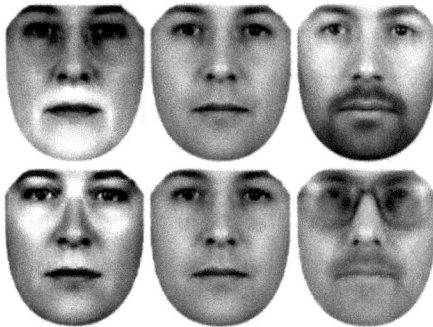

Figure 2.17 Examples of synthesized face images using AAM. Courtesy of the authors from Ref. [24].

pattern in δI provides information regarding how to adjust the model parameters. Toward this end, they model the mapping linearly, i.e., $\delta c = A\delta I$ and find A using multivariate regression with given sample of known δc and δI. The training set of δc and δI are generated in an offline process, where a known model parameter c is randomly perturbed to generate variations of I_m. The deltas of the model images and the model parameters are collected to form the training set to find A.

Given a new image and the mapping A, the matching process begins with the initial δI. It then updates the model parameter c with adjustment δc as predicted as $A\delta I$, followed by the generation of a new model image, from which δI is recomputed. The procedure repeats until δI is small enough that no improvement is possible for updating the model parameters.

Optimizing the AAM for Mobile Platforms
In order to make real-time tracking possible on mobile platforms, the authors make three further modifications [24]:

1. *Incremental parameter learning and pixel sampling*: In the original implementation, every image pixel is used to compute δI and every model parameter is updated. For further robustness and efficiency, the authors propose a sequential modeling approach: In the earlier stages, the algorithm samples fewer pixels from the image and predicts fewer model parameters while in later stages, more pixels are used to predict more model parameters. The effect is that coarse models are constructed using lower-resolution images and higher-resolution images are used to build finer-grained models to handle details. The motivation behind this is that it will not be effective to predict the details when the solution is still far from optimal. This also helps prevent the optimization procedure from falling into local minima in the early stages.
2. *Haar-like feature-based regression*: Instead of directly using image pixels, the authors propose to use rectangle features as in the face detector. Features are a more compact representation, reducing the number of required parameters. These Haar-like features can be computed efficiently using integral images, as introduced previously.
3. *Use of fixed-point operations*: Avoid using floating-point operations in the implementation reduces computational loads. The Haar feature and the integral images can be computed entirely on fixed-point operations.

The authors implement the entire facial feature tracking system on Nokia N900 and evaluate it with a public dataset containing about 600 training images and about 1,500 testing images, all with labeled landmarks. The accuracy of the system is quantified by measuring the point-to-point errors over all the landmark points. The system is able to achieve a frame rate of 50 frame per seconds and maintain a $<10\%$ error rate.

Applications

To demonstrate the applicability of the mobile AAM, the authors propose a mobile biometrics system [25]. The system provides a verification layer where the users could use their face and voice for remote login without typing username and password.

The system first determines the face appearance by detecting the face region. Instead of using the rectangle features, the LBP representation is used to summarize the image statistics for a candidate region. A cascade of classifiers is built based on the LPB features in order to reject false positives. Later, the AAM is used to localize individual facial features, such as eyes or nose. These facial features are used to remove irrelevant background. They are also used to fit the unseen face image into a predefined face shape model so that the scale, shape, and orientation of input image is normalized to that of those in the face database. The processed face image is then sent for face verification [19] after being normalized for lighting. Figure 2.18 illustrates this process.

Note that the authors could have tried to recognize the face using the detected face region without using AAM. However, the variations in face orientations, face expressions, lighting conditions, and background clutter could degrade the face recognition system. The accuracy of face recognition could be further improved when these factors are removed or reduced.

The system also uses voice recognition to improve the accuracy of authentication. As seen in Figure 2.19, the image data and voice data are fed separately into the system and processed independently. The face module takes and processes the image to give a score on how likely it is that a face belongs to the authentic users. The same goes for the voice module. In the data fusion step, another classifier takes the scores from both modules and determines whether to grant access.

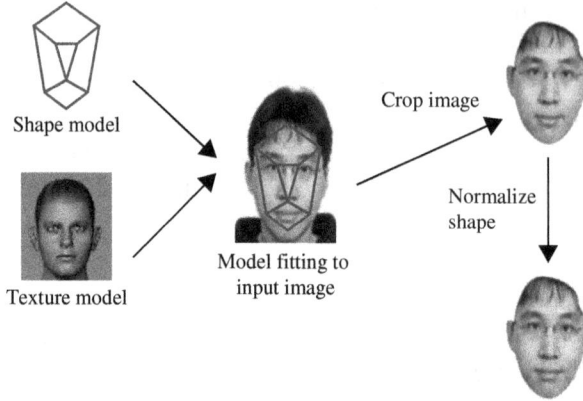

Figure 2.18 Illustration of using AAM to preprocess the detected face region. Reproduced from figure 3 in Ref. [25].
Given a detected face region, AAM algorithm is used to fit the best shape and texture model to the input image in
order to locate the facial features. Once located, the irrelevant information such as background or shirt is discarded to
generate a cleaner face image, which is later shape-normalized. The final face image is sent for verification.

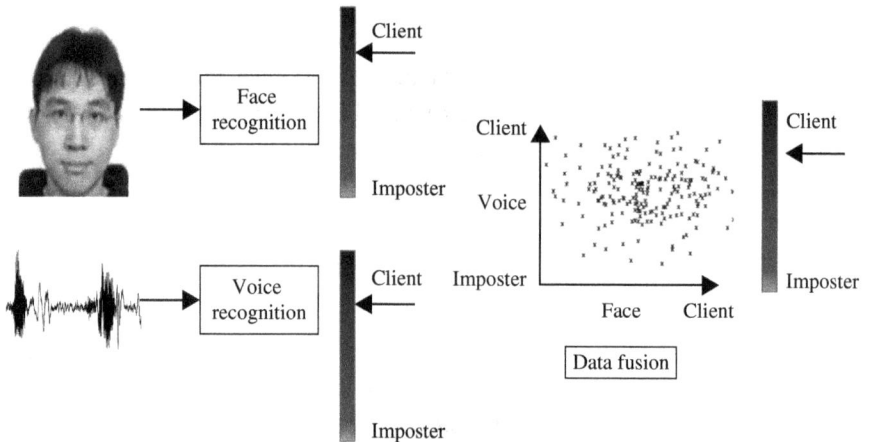

Figure 2.19 Illustration of the proposed system in Ref. [25]. Reproduced from figure 1 in Ref. [25]. Given a video
sequence containing voice and image data, the face-recognition module in the system gives a score on how likely it
is that the detected face belongs to the client. The same goes for the voice data. A data fusion module then makes
the final decision based on the scores from the two modules.

Other Applications on Mobile Platforms

The facial features are very useful in another application domain that
needs higher-level face analysis, such as face expression recognition.
For example, in Ref. [26], the authors present a system that uses sensor
data from a tablet to infer the reactions of the user when he/she is

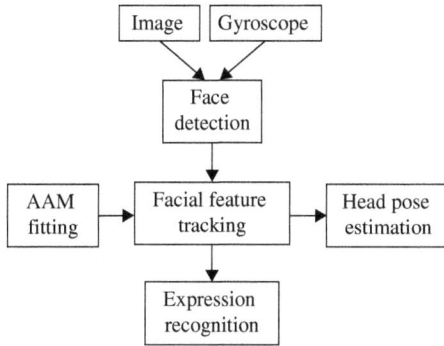

Figure 2.20 Illustration of the proposed system in Ref. [27]. Reproduced from figure 1 in Ref. [27].

watching movies. The inferred reactions are transformed to a movie rating automatically. One module inside their system is the face analysis system, which tracks the face, eyes, and lips. The locations and sizes of eyes and lips are used as features for user reaction recognition.

To cite another example, Visage system [27] tracks the face, detects the facial features, and uses the facial features to recognize user emotion, e.g., sad or happy. Two proof-of-concept applications are built on top of Visage to demonstrate its applicability. For instance, the Street View+ shows different viewpoints of Google Street View based on detected head poses, while the MoodProfile detects and records users' emotion when they use a particular application. Figure 2.20 illustrates the flow chart of the Visage system.

The Visage system starts with phone pose detection based on the sensor on the phone. The pose information is used to compensate for the tilting of the face images. Face detection based on the Viola–Jones detector is done to locate the face region. Later on, the AAM approach is used to detect the facial features on the face. Finally, these facial features are used to classify the user emotion into one of the seven major classes.

SUMMARY

In this chapter, we introduce and survey relevant face technologies. We start with face detection, as it is usually the inception of other face analysis algorithms, such as face recognition and facial feature tracking. We discuss the widely used and most successful face-detection algorithm—the Viola–Jones face detector. Given a region containing a face, we then give an overview of methods for face recognition. We focus on widely used face recognition algorithms—Eigenfaces and Fisherfaces—and briefly touch on methods from other genres.

We then discuss strategies for optimizing the performance of the Viola–Jones face detector for mobile applications, as it is usually the most time-consuming step in any mobile face-related applications. Based on the optimized face detectors, we introduce mobile applications that utilize face detection, such as using head pose estimation for new types of user interface. After that, we talk about the applications of face recognition in the domain of biometrics authentication where the mobile phone could unlock itself by recognizing the face of the user.

Following face recognition, we discuss the working principles and applications of facial feature tracking. In particular, we introduce the AAM and how it can be modified for mobile applications. We also illustrate an authentication application that uses AAM for better face recognition.

In summary, through this chapter, we can see that the main strategies for making efficient algorithms are as follows:

1. The use of fixed-point operations—avoiding floating-point data types reduces the amount of computation significantly. We see the principle at work for the Viola–Jones detector and the AAM algorithm.
2. Downsampling data and parameters. For example, downsampling the image sizes accelerates the Viola–Jones detector. Also, in the implementation of the AAM algorithm, predicting a subset of model parameters at a time improves not only the robustness but also the efficiency of the algorithm.
3. The use of application-specific information or other auxiliary sensors. For example, mobile applications typically need to deal with frontal faces only. The data coming from the gyroscope on the mobile phone could also be used to infer the phone pose and compensate for the tilt of the image, therefore reducing the need to deal with faces with different orientation and tilting.

REFERENCES

[1] Vaquero D, Feris R, Tran D, Brown L, Hampapur A, Turk, M. Attribute-based people search in surveillance environments. In: IEEE workshop on applications of computer vision; 2009.

[2] Thornton J, Baran-Gale J, Butler D, Chan M, Zwahlen, H. Person attribute search for large-area video surveillance. In: IEEE international conference on technologies for homeland security; 2011.

[3] Yang M-H, Kriegman D, Ahuja N. Detecting faces in images: a survey. IEEE Trans Pattern Anal Mach Intell 2002;24:34–58.

[4] Viola P, Jones M. Robust real-time object detection. Int J Comput Vis 2001;57(2):137–54.

[5] Freund Y, Schapire R. A decision-theoretic generalization of on-line learning and an application to boosting. J Comput Syst Sci 1997;55:119–39.

[6] Zhao W, Chellappa R, Phillips P, Rosenfeld A. Face recognition: a literature survey. ACM Comput Surv 2003;35:399–458.

[7] Turk M, Pentland A. Face recognition using Eigenfaces. In: IEEE international conference on computer vision and pattern recognition; 1991.

[8] Belhumeur P, Hespanha J, Kriegman D. Eigenfaces vs. Fisherfaces: recognition using class specific linear projection. IEEE Trans Pattern Anal Mach Intell 1997;19:711–20.

[9] Wiskott L, Fellous J-M, Krueger N, von der Malsburg C. Face recognition by elastic bunch graph matching. IEEE Trans Pattern Anal Mach Intell 1997;19:775–9.

[10] Pentland A, Moghaddam B, Starner T. View based and modular Eigenspaces for face recognition. In: IEEE conference on computer vision and pattern recognition; 1994.

[11] Ren J, Kehtarnavazl N, Estevez L. Real-Time optimization of Viola–Jones face detection for mobile platforms. In: IEEE Dallas circuits and systems workshop; 2009.

[12] Scheuermann B, Ehlers A, Riazy H, Baumann F, Rosenhahn B. Ego-motion compensated face detection on a mobile device. In: IEEE conference on computer vision and pattern recognition workshops (CVPRW); 2011.

[13] Rahman M, Kehtarnavaz N, Ren J. A hybrid face detection approach for real-time deployment on mobile devices. In: IEEE international conference on image processing; 2009.

[14] Wang Q, Wu J, Long C, Li B. P-FAD: real-time face detection scheme on embedded smart cameras. IEEE J Emerging Sel Top Circuits Syst 2013;3:210.

[15] Ren J, Rahman M, Kehtarnavaz N, Estevez L. Real-time head pose estimation on mobile platforms. J Syst Cybern Inf 2010;8:56.

[16] Hansen T, Eriksson E, Lykke-Olesen A. Use your head—exploring face tracking for mobile interaction. In: ACM SIGCHI; 2006.

[17] Wang Y-C, Donyanavard B, Cheng K-TT. Energy-aware real-time face recognition system on mobile CPU-GPU platform. In: European conference on computer vision workshop (ECCVW); 2010.

[18] Su Y, Shan S, Chen X, Gao W. Hierarchical ensemble of global and local classifiers for face recognition. IEEE Trans Image Process 2009;18.

[19] Hadid A, Heikkild JY, Silven O, Pietikdinen M. Face and eye detection for person authentication in mobile phones. In: ACM/IEEE international conference on distributed smart cameras; 2007.

[20] Xi K, Tang Y, Hu J, Han F. A correlation based face verification scheme designed for mobile device access control: from algorithm to Java ME implementation. In: IEEE conference on industrial electronics and applications (ICIEA); 2010.

[21] Ojala T, Pietikainen M, Maenpaa T. Multiresolution gray-scale and rotation invariant texture classification with local binary patterns. IEEE Trans Pattern Anal Mach Intell 2002;24:971−87.

[22] McCool C, Marcel S, Hadid A, Pietikainen M, Matejka P, Cernocky J, et al. Bi-modal person recognition on a mobile phone: using mobile phone data. In: IEEE international conference on multimedia and expo workshops (ICMEW); 2012.

[23] Cootes T, Edwards G, Taylor C. Active appearance models. In: European conference on computer vision; 1998.

[24] Tresadern PA, Ionita MC, Cootes TF. Real-time facial feature tracking on a mobile device. Int J Comput Vis 2012;96(3):280−9.

[25] Tresadern P, Cootes T, Poh N, Matejka P, Hadid A, Levy C, et al. Mobile biometrics: combined face and voice verification for a mobile platform. IEEE Pervasive 2013;12:79−87.

[26] Bao X, Fan S, Varshavsky A, Li K, Choudhury R. Your reactions suggest you liked the movie: automatic content rating via reaction sensing. In: ACM UbiComp; 2013.

[27] Yang X, You C-W, Lu H, Lin M, Lane N, Campbell A. Visage: a face interpretation engine for smartphone applications. In: International conference on mobile computing, applications, and services; 2012.

www.ingramcontent.com/pod-product-compliance
Lightning Source LLC
Chambersburg PA
CBHW060514220326
41598CB00025B/3655